# BEHIND THE SCENES

A TIM OLK GALLERY

# BEHIND THE SCENES

## A TIM OLK GALLERY

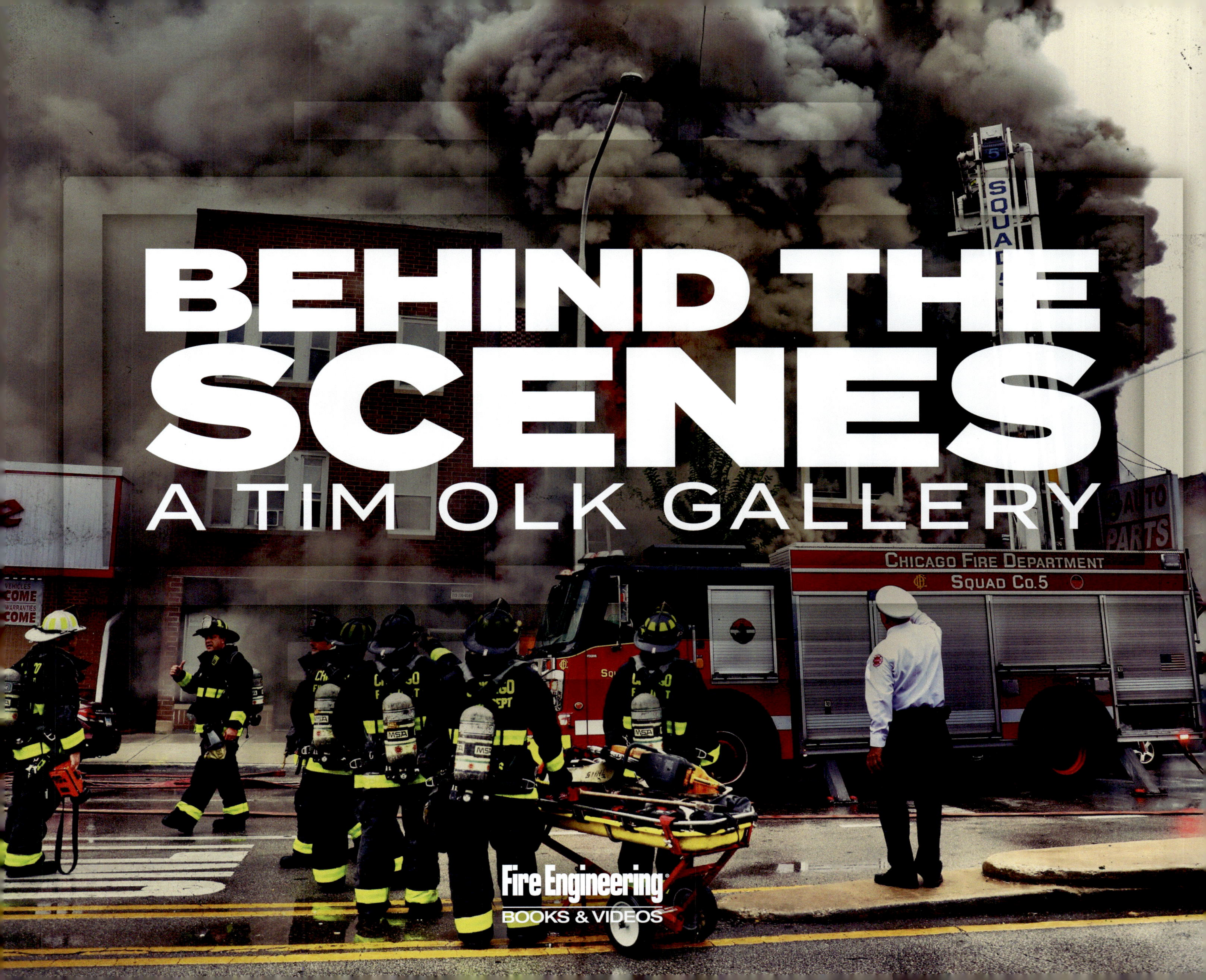


Fire Engineering
BOOKS & VIDEOS

**Disclaimer**

The recommendations, advice, descriptions, and methods in this book are presented solely for educational purposes. The author and publisher assume no liability whatsoever for any loss or damage that results from the use of any of the material in this book. Use of the material in this book is solely at the risk of the user.


Fire Engineering Books & Videos
110 S. Hartford Ave., Suite 200
Tulsa, Oklahoma 74120 USA

800.752.9764
+1.918.831.9421
info@fireengineeringbooks.com
www.FireEngineeringBooks.com

Executive Vice President: Eric Schlett
Vice President, Group Publishing: Amanda Champion
Acquisitions: David Rhodes, Diane Rothschild
Director, eLearning and Books: Starlet Franz
Production Manager: Tony Quinn
Developmental Editor: Chris Barton
Operations Manager: Holly Fournier
Book and Cover Designer: Brandon Ash

Library of Congress cataloging-in-publication data available on request.



Printed in the United States of America

1 2 3 4 5  26 25 24 23

# DEDICATION

This book is dedicated to all my friends in the fire service who have passed away.

**Jerry Olk**, my dad
**Bobby Halton**, *Fire Engineering*
**Wayne Luecht**, Northbrook Fire Department
**Frank Nunez**, Itasca Fire Protection District
**George**, Indianapolis Fire Department fan
**Mike Higgins**, Grayslake Fire Department
**Kevin Fitzpatrick**, Chicago Tower 34
**Bradley Liggett**, chief, Freeport Fire
**Keith A. Peacy**, North Chicago Fire
**Lt. Eric Hosette**, Clinton, IA, Fire Department
**Tom Bauer**, Des Plaines Fire Department
**Dean Schroeder**, Deerfield Fire Department
**Michael Schubert**, Chicago Fire Department Engine 75
**Joseph Pavlik**, Justice Police Department
**Cathy Wagner**, Rochester Volunteer Fire Department
**Albert W. Schlick III**, Huntley Fire Protection District
**Dan Capuano**, Chicago Fire Department Tower Ladder 34
**Dylan Cunningham**, South Holland Fire Department
**Colton Gritzmacher**, Boone County Fire District 2
**Mark Amore**, Lincolnshire Riverwoods Fire
**Gerald Ellis**, Illinois State Trooper
**Ed Dagdick**, Wauconda Fire Department
**Herbie Johnson**, Chicago Fire Department Engine 123
**Chris Wheatley**, Chicago Fire Department Truck 2
**William Gitzke**, Cary Fire Protection District
**Pat Hannon**, Chicago Fire Department Truck 8
**Juan Bucio**, Chicago Fire Department
**Jayson Mayoski**, Chicago Fire Department
**Ed Stringer**, Chicago Fire Department Engine 63
**Corey Ankum**, Chicago Fire Department Tower Ladder 34
**John Leonard**, Chicago Fire Department Truck 20
**Paul Miller**, Desplaines Fire Department
**David Jones**, Evanston Fire Department chaplin
**Russ Constantino**, Gurnee Fire Department
**John "Mac" McCastland**
**Bernie Schmidt**, Northbrook Fire Department
**Analise Chapman**, Riverdale, IA, Fire Department
**Cory Barr**, Sun Prairie Fire Department
**Les Macko**, fire chief, Cary Fire Protection District (ret.)

# ACKNOWLEDGMENTS


Thank you to all of the people who have taught me over the years and to those I call my friends.

**Bobby Hoff**, Chicago Fire Department
**Gordon Nord Jr.**, Lyons Fire Department
**Tom Olk**, captain, Northfield Fire Rescue
**Bill Muno**, Evanston Fire Department
**Ray Orozco Jr.**, Chicago Fire Department
**Bob Brogan**, Northbrook Fire Department (ret.)
**Jim Hoover**, Northbrook Fire Department (ret.)
**Ray Larson**, Deerfield Fire Protection District
**Mike Agostinelli**, Chicago Tower Ladder 14
**Art Aranda**, Chicago Fire Department Truck 41
**Jim Arie**, chief, Barrington Fire Department (ret.)
**Jenny Bahr**, Itasca Fire Protection District
**Mike Dugan**, FDNY
**Jack Barry**, Franklin Park Fire Department (ret.)
**Kelly Beagley**, Chicago Fire Department
**Chicago Fire Department**
**Frank Bertog**, Chicago Fire Department (ret.)
**Tim Sampey**, Chicago Fire Department (ret.)
**Mike Burns**, Chicago Fire Department (ret.)
**Keith Vallow**, Harvey Fire Department
**Brian Mckermitt**, Chicago Fire Department (ret.)
**Thomas Vogenthaler**, Chicago Fire Department
**Mitch Crooker**, Chicago Fire Department (ret.)
**Larry McCormack**, Chicago Fire Department Squad 5
**Richard Mikel**, Elk Grove Fire Department
**Pat Kehoe**, Chicago Fire Department (ret.)
**Steve Chikerotis**, Chicago Fire Department
**Rick Lasky**, Fire Engineering
**John Salka**, FDNY
**Northbrook Fire Department**
**Glenview Fire Department**
**Matt Knott**, fire chief, Green Bay Fire Department
**Highland Park Fire Department**
**Joseph Schrage**, fire chief, Highland Park
**Ron and Janeen**, Joynt Decal Works
**Talon Joynt**, my little brother
**Dave Morhy**, Northbrook Fire Department
**Barb and Rory Ohse**, Chicago Fire Department
**Kelly Burns**, Chicago Fire Department
**Brian Kunce**, Boone County Fire District 2
**Dennis Dovenmuehle**, Boone County Fire District
**Jake Jakubec**, Chicago Battalion 6
**Don Heroma**, Chicago Battalion 6 (ret.)

# CONTENTS

# ABOUT THE PHOTOGRAPHER

**Tim Olk** took his first fire scene photos when he was 13 years old. He brought the pictures into the fire station, where he met Northbrook Assistant Chief and Fire Investigator Wayne Luecht. He took photography classes and learned how to shoot fire scenes from Luecht. In 1999, Tim lost his mentor and friend when Chief Luecht died of injuries sustained on duty.

Eventually, Olk became a paid-on-call firefighter and was certified in advanced fire scene photography by the Fire Investigators Strike Force (FISF), a nonprofit association of fire investigation professionals dedicated to the education of fire investigators.

A photographer for nearly 40 years, Olk was a paid-on-call firefighters for Northbrook Prospect Heights, IL, and Northfield, IL. He is also the photo unit photographer for Boone County, IL, Fire District 2 and Des Plaines, IL, Fire Department.

Olk travels all over the greater Chicago area capturing scenes and incidents like the many shared in this book

Dozens of his photos have been featured in *Fire Engineering* and *Fire Rescue* magazines.

CHICAGO
FIRE DEPT.
BATTALION
CHIEF
HEADQUARTERS
CHICAGO
OLK

CHICAGO
FIRE DEPT.

CHICAGO
FIRE DEPT.

ONE WAY
Engineering
FDIC

# BUILDING FIRES

A building fire is a fire involving a structure of various types from commercial to industrial buildings.

Residential buildings can be single-family detached homes and townhouses to apartments and tower blocks, or various commercial buildings ranging from offices to shopping malls. This is in contrast to "room and contents" fires, chimney fires, vehicle fires, wildfires, or other outdoor fires.

Structure fires typically have a similar response from the fire department that includes engines, ladder trucks, rescue squads, chief officers, and an EMS unit, each of which will have specific initial assignments. The actual response and assignments will vary between fire departments. It is not unusual for some fire departments to have a predetermined mobilization plan for when a fire incident is reported in certain structures in their area. This plan may include mobilizing the nearest aerial firefighting vehicle to a tower block or a foam-carrying vehicle to structures known to contain certain hazardous chemicals.

In June 2021, a massive fire and hazmat incident destroyed Chemtool Inc. in Rockton, IL, and forced home evacuations within a mile of the incident. Olk took more than 6,000 photos of the blaze—several of which are included here—that burned for four days and took two days to contain.

Janesville, IL, Fire Department Tower Ladder 81 works a Rockton chemical plant fire.

Chicago Fire Department Squad Co #2 works an extra alarm at an auto body warehouse fire.

Janesville, IL, Fire Department firefighters work on the Tower Ladder 81 basket at a Rockton chemical plant fire.

Chicago Fire Department Lt. Art Aranda gives orders to Squad 2 firefighters.

The *Chicago Fire* television program films a scene on Chicago's West Side.

The American flag flies at the scene of a major chemical plant fire in Rockton, IL.

Chicago Fire Department Engine 38 works in front of a building at an auto body warehouse fire.

Rockford Fire Department Ladder 1 works a major factory fire on the northwest side.

Chicago firefighters respond to an extra alarm at apartments over a storefront.

Elmwood Park Fire Department Truck 945 at an extra alarm at an Oak Park fire.

Chicago Fire Department Engine 109 works a junkyard fire on the West Sde.

Chicago Fire Department Tower Ladder 37 and Truck 15 firefighters work a roof fire at 45th and Ellis.

A Chicago Fire Department candidate firefighter clears out a second-floor window.

Lake Forest, IL, Fire Department Tower Ladder 421 works an extra-alarm fire in Becch Park.

Chicago Fire Department Squad Co #2 works at an extra-alarm fire at 2943 N. Milwaukee Ave.

Chicago firefighters set up defensive operations at an extra-alarm fire on south Michigan Avenue.

Multiple tankers line up to drop water at a major chemical plant fire In Rockton, IL.

Boone County Fire District 2 Ladder 357 works on the southeast side of building with multiple explosions.

Boone County Fire District 2 Ladder 357 prepares to set up on the southeast side of a building at a Rockton, IL, chemical plant fire.

Engine 813 supplies water to several handlines at a major chemical plant fire.

The Wind Lake, WI, Fire Department responds to a large fire at a recycling plant.

A chemical plant fire at Rockton, IL.

Kenosha, IL, Fire Department Truck Co #7 at an extra-alarm fire at 22nd Avenue and 62nd.

Delevan, IL, Fire Department Engine 5 is drafting at a chemical plant fire at Rockton, IL.

A chemical plant fire at Rockton, IL.

Janesville, IL, Tower Ladder 81

Chicago Fire Department Engine 46

Forestview, IL, engine feeds Central Stickney, IL, tower ladder at a large rubbish fire.

Chicago Fire Department Squad Co. #7 at an extra-alarm fire in Bensenville, IL.

Rockford Fire Department Truck 1

Chicago Fire Department still-and-box alarm apartment fire on Chicago's Southside.

Chicago Fire Department Aerial Tower 1 at an extra-alarm fire at a flea market.

Chicago Fire Department Squad Co. #5 at an extra-alarm fire on South Michigan Avenue.

Rockford firefighters work at an extra-alarm factory fire.

Chicago Fire Department Squad Co. #7 at a large apartment fire in Bensenville, IL.

Chicago Fire Department Squad Co. #5

Northfield, IL, Fire Captain Tom Olk

# HOUSE FIRES

As more people live closer and closer together, particularly as a result of urbanization, risks associated with fire increase. House fires are predominantly accidental, usually involving indoor lamps, stoves, or heaters that use open flames. Careless use of cigarettes is another leading cause of fires, and fires may also be set deliberately as public violence and arson.

House fires affect hundreds of thousands of homes per year around the world. Fire spreads quickly and can easily engulf a home and neighboring structures within minutes. In the city, fire risk is often aggravated by the difficulty of access for fire response equipment into narrow streets, and high-rise buildings present another challenge when the middle and top floors of the building may only be accessible by stairs.

On the northwest side of Chicago, Cary, IL, firefighters battle a garage fire.

Firefighters went to the second floor to fight flames on March 14, 2020, when an interior stairwell was burned out.

A Truck 27 firefighter attempts to complete roof ventilation and was ordered off the roof when fire appeared on June 30, 2019.

Chicago firefighters Art Aranda (right) and Billy Heenan of Squad 5 attack fire from a second-floor landing because the interior stairs were burned away on March 14, 2020.

Firefighters draft water in a non-hydrant area.

Chicago firefighters save one of their own during a Mayday incident when a firefighter became trapped inside on Jan. 4, 2020.

Chicago firefighters enter a building in an attempt to rescue two people reportedly trapped inside on Nov. 22, 2020.

Chicago Fire Department Engine 117 and Tower Ladder 14 in an alley at a garage fire on Aug 12, 2019.

Chicago firefighters perfor n overhaul after fighting a working apartment fire.

Chicago firefighters located a body during a 2-11 alarm fire on a second-floor house on Jan. 7, 2017. An interior stairwell was burned out, so retrieval happened outside the window, and the deceased were placed in a Stokes basket on Tower Ladder 14.

Raymond Cullar, a Chicago firefighter from Squad 5, uses a halligan tool to take out a window frame.

Chicago firefighters enter a building to try to rescue people who may be trapped inside on Nov. 11, 2020.

Some type of gas explosion damaged other houses in this neighborhood on Jan. 16, 2019, where the Northbrook, IL, Fire Department responded to a 2-11 alarm house fire.

Chicago firefighters get down off a roof after ventilation on April 4, 2018.

Chicago Fire Department responded to a 2-11 alarm blaze with exposures to two other houses. Two truck company firefighters are going to the roof to vent on March 19, 2017.

Chicago Squad 5 firefighters at the scene of smoke coming out of a second-story window.

Chicago firefighters rescue a woman from a house fire while crews vent the roof on May 27, 2021.

A Chicago Mayday; a roof fell on a firefighter.

Chicago Fire Department Squad 2 works at a vacant apartment building fire trying to open metal coverings on a window to gain access inside on Oct. 31, 2019.

Chicago firefighters rescue a fellow firefighter from Engine 82 on Jan. 4, 2020.

Chicago FD Squad 5 fights fire from the outside after air horns sounded, alerting everyone inside to evacuate on April 4, 2018.

Chicago firefighters deploy off the roof after successful ventilation on April 4, 2018.

Evanston, IL, firefighters ventilate the roof from a ladder truck on Jan. 9, 2015, where a Mayday occurred.

Heavy smoke shows for Chicago Squad 5 venting a roof at a southside house fire on June 21, 2020.

A West Burlington, IL, fire engine parked at a rural fire assisted by the Danville, IL, Fire Department.

Chicago firefighters work safely to suppress fire on the outside wall at an electrical box.

A burn down of a building after training exercises were completed in the town of Randall, WI, on Oct. 7, 2018. This is a "stairway to hell" view that was shot from a basement entrance of a fully involved two-story house—looking up to the first floor. Burn downs like these occur after all the training is completed and makes for less debris that the contractor needs to haul away.

# FUNERALS

Firefighters are often the unsung heroes in our communities. While the city sleeps, they are on-call and ready when a fire strikes. They sacrifice a lot to do this job, risking their lives every day. Whether it's a tragedy that strikes or a minor fire, they are usually the first to arrive on a scene. With dignity and grace, these men and women serve in a capacity that many can't even imagine. When we lose a firefighter to death, we should show them the honor and respect they deserve.

When we lose a firefighter, it is a loss to our entire community. There are special circumstances in place to honor a service member depending on the cause of death.
In some cases, a service member may qualify for a fire department funeral. The possibilities include the following:

- Line-of-duty death: This can include a traumatic death as a result of their service or a gradual decline in health postfire.
- Off-duty death: This can include any traumatic incident that occurs unrelated to work. It can also sometimes include a gradual decline in health due to a life-threatening event.
- Retired or former member: This may include previous firefighters and volunteer firefighters and can include a gradual decline in health or sudden death.

Colton Gritzmacher, a firefighter at Boone County, IL, Fire District 2, was killed in collision a few weeks after this photo was taken at a fire scene.

Boone County Fire District 2 members attend Colton's funeral.

"When I reflect on the photos taken at Colton's visitation and funeral service, I struggle to find the right words. *Grateful* is what comes to mind. When you are thrown into the mix of shock having just lost your loved one, and then having to make the decisions with final arrangements, your mind is completely overloaded. The firehouse becomes your lifeline. Anything you need and I mean *anything*, it was managed for you. I wanted to be everywhere, as I did not want to miss a thing with the procession of bringing Colton home to all the details that went into the planning and execution of his funeral service with firefighter burial rights. You would think that the last thing you want to do is to look at pictures of your loved one's funeral service, but I can assure you that it is one of my favorite things to do. The photos reflect the true meaning of the "brotherhood" that Colton was honored to be a part of, and they allowed me to see all the moments I was not able to see personally. The funeral service was the final gift I could give to my son, and having the photos allows me to look back with great gratitude.

—**Charmaine**, mother of Colton G.

Chicago area firefighters attended the funeral service for United States Marine Lance Corporal Peter Knobel to show their respect on Nov. 19, 2022.

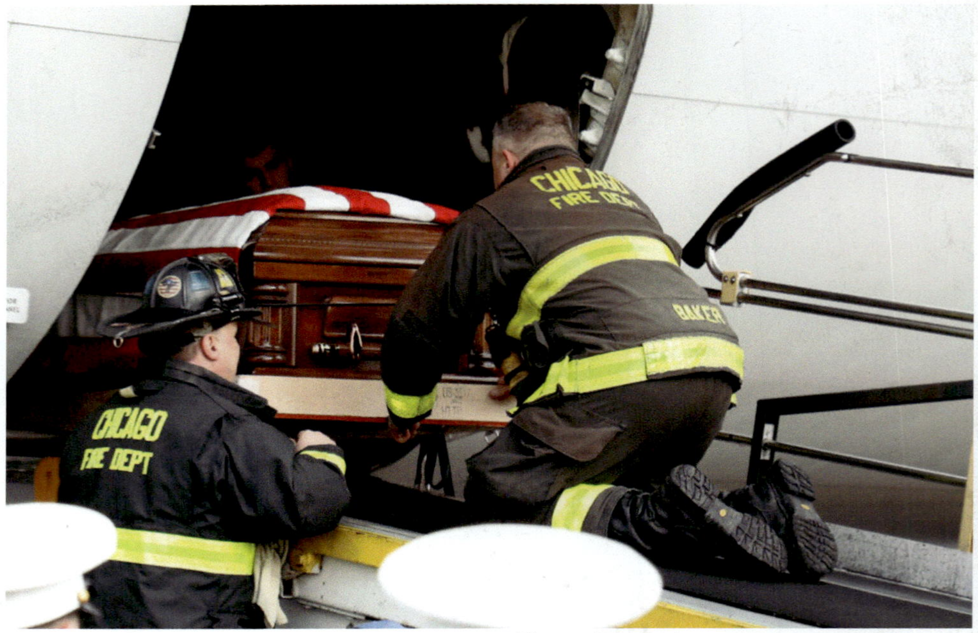


His remains were transported overseas to Chicago.

Chicago firefighters from Squad 7 prepare the remains of United States Marine Lance Corporal Peter Knobel upon arrival at O'Hare International Airport for his final ride home.

U.S. Marine gun salute at Knobel's funeral.

The American flag folded by Marines near the casket at Pete Knobel's service.

Honor guard with casket detail for Geneva, IL, Fire Department Battalion Chief (Ret.) Gavin Hayes, who died of cancer on Oct. 9, 2017.

Honor guard with casket detail for Geneva, IL, Fire Department Battalion Chief (Ret.) Gavin Hayes, who died of cancer on Oct. 9, 2017.

Grayslake, IL, Chief Higgins's gear on the fire truck

Grayslake, IL, Fire Protection District trustee Greg Keoppen at the casket at the funeral of Mike Higgins.

This photo was taken from a flat roof on the second story at St. Rita Catholic Church at the Chicago funeral of Dan Capuano.

Chicago firefighter Dan Capuano of Tower Ladder 34 died in the line of duty. His helmet and jacket are displayed at the front of the church on his rig.

Chicago Fire Department honor guard and pipes and drums perform on Dec. 18, 2015, at Dan Capuano's service.

The funeral service for Jim "Franco" Franciskovich's service was held March 2, 2017, in West Burlington, IA.

A photographer's vantage point from a hill looking down on firefighters unloading the casket of Jim "Franco" Franciskovich.

Comstock, MI, Township Fire Chief Ed Switalski was killed in a collision in Comstock Township. Switalski's wife called Olk, who knew the chief because he had worked in the Chicago area.

*Olk receives calls to photograph services from those in the greater Chicago area and surrounding states. He is welcomed to stay the night at local firehouses. He also frames a photograph of that first responder—if he has one—for use at the church, funeral home, or gravesite.*

A 1937 Seagrave fire truck from the Detroit Fire Department carried the casket of Ed Switalski on June 19, 2017.

Lt. Eric Hosette "Hoss" of Clinton, IA, was buried on Jan. 10, 2019. He was killed fighting a grain bin fire when there was an explosion.

Lt. Eric Hossette's funeral service at Clinton, IA.

Hossette's fire department with his wife, a fire department dispatcher, and his daughter.

Nine months later, Eric Hossette's daughter Addy Hosette helped dedicate a truck in his name on Oct. 23, 2019.

A Union Grove-Yorkville, WI, Fire Department truck transported Brian Serdynaski to his final resting place on Nov. 10, 2019. He died of a heart attack in the line of duty on Nov. 2, 2019.

*Funerals—like fires—don't always happen on nice, sunny days. Olk once photographed a funeral procession in a blizzard when firefighters carried the casket a few blocks from the funeral home to the outdoor auditorium. "Sometimes everyone is standing in 103-degree heat or it's freezing."*

Des Plaines, IL, Fire Department Deputy Chief Sam Foster brings a memorial flag to the Tom Bauer family on Oct. 15, 2017.

Tom Bauer's helmet is placed near the casket in the hearse, which is driven to the gravesite where the family will receive it.

Sam Foster offers the memorial flag to the Bauer family.

*Olk often frames photos of those to be remembered such as this photo of Bauer at a Des Plaines graveside service.*

*"I am giving back" to the fire service, explains Olk, who prioritizes his photography with attending funerals of first responders lost in the line of duty, from cancer or those who have retired from service.*

Ed Dagdick's daughter holds his dress cap. A fire instructor, he got cancer within months of Olk taking his retirement photo. He died on March 30, 2019. He was deputy chief of Wauconda, IL, Fire District.

Ed Dagdick with his grandkids in a retirement photo taken on Dec. 28, 2018, three months before his death.

Funeral services for East Joilet, IL, Fire Protection District Fire Chief Robert F. Scholtes Jr. were held on Dec. 30, 2019, four days after he passed away.

Evanston, IL, Fire Rescue Department Chaplain David Jones's services were held on Feb. 21, 2020.

A bagpiper at the graveside funeral services of Itasca, IL, firefighter Frankie Nunez, who died of cancer.

Services for Chicago police officer Marco DiFranco on April 9, 2020, five days after he died of COVID-19.

South Holland, IL, Firefighter Dylan Cunningham's funeral service was Oct. 6, 2020. He died in the line of duty.

Funeral services were held on January 13, 2021, for Huntley Fire Protection District Deputy Chief Albert W. Schlick III, who was also an instructor at FDIC.

Schaumberg, IL, Firefighter Mike Solberg died of brain cancer on May 6, 2016.

This fire engine was detailed for Lt. Jim Carney of Lincolnshire Riverwoods, IL, Fire Protection District at the firehouse before his funeral on Dec. 21, 2017.

Lt. Artie Aranda of the Chicago Fire Department honor guard has attended funerals where he presents the City of Chicago flag to the family.

On April 13, 2020, firefighters unloaded the casket of Truck Co. 25 Firefighter Mario Araujo.

Chicago Firefighter Edward Singleton of Engine 127 was buried on April 22, 2020.

A photo of a flag at the services for Riverdale, IA, Firefighter Analiese Chapman on Feb. 7, 2020. She died from cystic fibrosis.

Honor guard and casket detail for Analiese Chapman funeral services

Firefighters line up outside St. Rita of Cascia Catholic Church for the arrival of Captain Herbie Johnson on Nov. 8, 2012.

Chris Wheatley's funeral procession from the vantage point atop the roof of the funeral home.

Chicago Fire Department Captain Herbie Johnson's funeral view taken from an engine.

*Olk's job priorities are funerals for first responders, so he will ask fellow photographers to cover emergencies in the Chicago area.*

*He arrives hours before funeral services to survey the route and distance between the funeral home or church. He arrives at the cemetery ahead of the funeral procession or hours before the actual service to photograph the memorialized gravesite decorated with pictures, flowers, and flags.*

# FIRE CHIEFS

A fire chief or fire commissioner is a top executive rank or commanding officer in a fire department. During an emergency incident, the first fire officer on the scene must "establish command," which can then be transferred to more senior officers, such as the chief. The chief may delegate some statutory powers to qualified officers, such as the ability to enter or use private property as reasonably necessary to stop a fire, or to order people or property seized as may be essential to preserving safety or investigating the cause of an incident.

Chicago Fire Department Fire Commissioner (Ret.) Ray Orozco

Chicago Fire Department Fire Commissioner (Ret.) Bob Hoff

Chicago Fire Department Battalion 19 Chief (Ret.) Mitch Crooker celebrated 40 years on the job in 2020.

Chicago Fire Deptment Battalion Chief (Ret.) Mitch Crooker; Father Tom Mulcrone, chaplain for almost 30 years; and District Chief (Ret.) Tom Carbonneau, who served 42 years with the department

Chicago Fire Department 19 Battalion Chief (Ret.) Mitch Crooker (left) and 18th Battalion Chief (Ret.) Mike Burns

Chicago Fire Department Battallion Chief (Ret.) Mitch Crooker and Deputy District Chief Anthony "Kool-Aid" Frazier

Chicago Fire Department Chief (Ret.) Tim Sampey

Chicago Fire Department 9th Battalion Chief (Ret.) Robert Sampey
with 18th Battalion Chief (Ret.) Tim Sampey

*"Tim Olk has done more for the morale of Chicago firefighters than anyone else," says* ***Retired Battalion Fire Chief Tim Sampey.*** *"His photos capture the heart of the Chicago Fire Department and many line the hallway at the CFD headquarters and in many firehouses."*

*"Olk covers a lot of ground within Chicago's 32 square miles from O'Hare Airport to the Indiana state border. He shows up at planned and unplanned events. He prioritizes attending first responder memorial services, funerals and wakes."*

*"The art he does is big as himself," Sampey says. "He is courteous and knows the traditions and doesn't step over the line. He doesn't ask for anything and treats everyone with professionalism, no matter their rank."*

Chicago Fire Department District Chief (Ret.) Richard Edgeworth

Chicago Fire Department Battalion Chief (Ret.) Richard A. Dory

Chicago Fire Department Special Operations Dive Chief 686 Jason Loch

Chicago Fire Department Deputy District Chief Bob Jurewicz

Chicago Fire Department Battalion 6 Chief John "Jake" Jakubec

Chicago Fire Department Battalion 10 Chief (Ret.) Brian McKermitt

Fire Commissioner Annette Nance-Holt is a 31-year veteran of the Chicago Fire Department who is responsible for the day-to-day operational and administrative oversight of the second-largest fire department in the United States.

Chicago Fire Department 18th Battalion Chief (Ret.) Ken Wojtecki

Harlem-Roscoe Fire Protection District Captain Mike Huffman is an Illinois State Certified Fire Instructor and received a Certificate of Honor from the Office of Illinois State Fire Marshal for helping save a young girl from a burning home while working for the North Park, IL, Fire Department.

Richton Park, IL, Fire Chief Mick Smith, former Village of Riverdale, IL, fire chief and youngest-ever fire chief at Phoenix, IL, Fire Department (left) with Elk Grove Village Fire Chief Richard Mikel, the 2018 Illinois Fire Chief of the Year

West Chicago Fire Chief Pat Tanner

Dalton Fire Department Fire Chief Pete McCain

Newport Township, IL, Fire Protection District Fire Chief Mark Kirschhoffer, a former Illinois Fire Chief of the Year

North Maine, IL, Fire Protection District Chief (Ret.) Mike Bronk

North Maine, IL, Fire Protection District Chief (Ret.) Mike Fox

Kansasville, WI, Fire Rescue Chief (Ret.) Scott Remer

Lyons, IL, Fire Department Fire Chief Gordon Nord Jr.

From left: Lyons, IL,Fire Chief Gordon Nord Jr.; Chicago Fire Commissioner (Ret.) Richard Ford; and Chicago Fire Department Deputy District 2 Chief Dave Dietz

Boone County, IL, Fire District 2 Deputy Chief Cory Dovenmuehle

Boone County, IL, Fire District Deputy Fire Chief Dennis Dovenmuehle

Franklin Park, IL, Fire Departmenmt Fire Chief (Ret.) Dave Traiforos

Escambia County, FL, Fire Rescue Battalion Chief Curt Isakson

Evanston, IL, Fire Department Deputy Fire Chief Billy Muno

Chicago Fire Department District Chief Walter Schroeder

Illinois Fire Chiefs Association 2023 Board of Directors

Roselle Deputy Fire Chief Tim Smeltzer and Long Grove, IL, Deputy Chief Andy Wienckowski

Itasca, IL, Fire Protection District Fire Chief John Schneidwind

(From left) Chicago Fire Department
Deputy District Chief (Ret.) Steve Chikerotis with Chief (Ret.) Bobby Halton

# WHAT THEY ARE SAYING

> “I have seen Tim’s compassion and wisdom, required of a professional photographer when taking photos at highly emotional scenes. Tim understands how to be empathetic and to see the situation through the eyes of those he’s trying to capture in his photos.
>
> **—Bobby Halton**

# COLLISIONS, PIN-INS

A pin-in car collision traps one or more occupants inside a vehicle. Pin-in car incidents are often serious, resulting in severe injuries or fatalities. Knowing the facts about pin-in car collisions can help drivers prevent them from happening.

In order for a pin-in car accident to happen, the impact of the crash must be great enough to either flip the car over or smash it into another vehicle or object hard enough to compress the body of the car. This makes a pin-in car accident quite different from a common fender-bender. Examples of pin-in car accidents include the following:

- A speeding car flips over, and the roof is compressed into the body.
- A car runs off the road and rolls down an embankment, crushing the doors so it cannot open.
- A drunk driver hits a car, pinning it against a guardrail.
- A large truck rear-ends a car, crushing it against the car in front.
- Two cars collide head-on, compressing the dashboards against the occupants so they cannot get out.

Boone County, IL, Fire District 2 handles a multiple vehicle collision where one person was trapped on May 20, 2017.

North Chicago Fire Department responded to a trapped person in a car where they worked for more than 90 minutes to extricate the victim. Flight For Life arrived at the scene of a car and dump truck collision where the person survived on Sept. 21, 2015.

The Chicago Fire Department deals with the aftermath of a car vs. garbage truck collision that killed three people on Oct. 15, 2013.

This garbage truck drove over a car, trapping three people, and burst into flames on Oct. 15, 2013.

Mundelein, IL, Fire Department uses a saw after a hillside rollover incident on an excavator with a drilling machine on Aug. 16, 2016.

Chicago firefighters work to free a person from a train vs. car collision after the car attempted to drive around the gates on Nov. 28, 2010.

The person extricated from this car after a train collision did not survive.

Chicago Fire Department Engine 106 stands by after a truck crashed over an expressway onto a road below. The driver survived the crash on March 25, 2017.

Glenview, IL, firefighters cut the roof off a vehicle to free an elderly couple on June 3, 2015.

Glenview, IL, firefighters stablilize a vehicle before cutting into it to free occupants.

Glenview, IL, Fire Department responds to a rollover crash to extricate multiple people on Dec. 18, 2011. Everyone survived.

Glenview, IL, Fire Department cleans up following a four-car collision with multiple extrications on Dec. 18, 2011.

Libertyville, IL, Fire Department works to free a trapped passenger after a one-car crash. At left, Newport, IL, Fire Chief Mark Kirchhoffer, who drove past the incident and stopped, watches as firefighters cut the door, on Nov. 6, 2005.

Libertyville, IL, Fire Department responds to a pin-in incident on Nov. 6, 2005, where Flight for Life lands at the scene.

Lincolnshire firefighters work to free victims from a plane crash on Nov. 28, 2011.

Lincolnshire Riverwoods, IL, firefighters free several victims from a medical transport plane crash in the woods. Three people perished in the crash, and a co-pilot and flight paramedic survived on Nov. 28, 2011.

Chicago Fire Department Squad Company 7 assists Lyons, IL, Fire Department to control a leaking tanker truck on April 21, 2019.

Lyons, IL, Fire Department requested assistance when a car went under a tanker truck hauling gasoline that leaked, so crews from O'Hare International Airport assisted with foam on April 21, 2019.

Lyons, IL, Fire Chief Gordon Nord Jr. directs where he wants foam applied at a rig crash on April 21, 2019.

Newport Fire Department responds to a fatality involving a jeep and a school bus where bus riders did not have serious injuries on April 5, 2013.

Northbrook, IL, firefighters work to extricate someone from a two-car collision. One person was transported to a medical helicopter on July 3, 2006, in this fatal incident.

Northbrook, IL, firefighters work to free a motorcycle driver who crashed through a front windshield of a car. Everyone involved survived this collision on May 20, 2016.

Four semi trucks and two cars were involved in a crash where Loyola Medical Flight transported a truck driver on June 4, 2008. Northbrook, IL, Fire Department responded to cut a door off a vehicle to free a driver. No serious injuries were reported.

Northbrock, IL, firefighters us the Jaws of Life® to extricate two people from a vehicle on Aug. 1, 2019.

Northbrook, IL, firefighters use Hurst tool cutters to free someone from a car that rolled over on the expressway in a marshy area on Aug. 8, 2020.

Northbrook, IL, firefighters remove a car door to free two people trapped in a two-car crash in which both survived on Aug. 1, 2019.

Yellow mustard can be seen coming from a truck at this crash scene involving four big rigs and two cars. Northbrook, IL, Fire Department responded to the scene on June 4, 2008.

Northfield, IL, Fire Rescue frees two people from a two-car collision on Dec. 26, 2012.

Wheeling, IL, firefighters work to free an occupant from a two-car crash with Flight For Life landing on July 3, 2006.

Downers Grove, IL, Fire Department responds to a collision on July 18, 2008.

Downers Grove, IL, Fire Department responds to a person trapped in a car after a three-vehicle crash.

# SPECIAL OPERATIONS

Fire department-based special operations personnel are dedicated to perform "heavy rescue" or "squad" duties. Agencies are responsible for establishing, staffing, operating, and maintaining special operations units and assigning and training those officers and firefighters.

According to author and FDNY Deputy Assistant Chief (Ret.) John Norman, special operations and "rescue companies can be vital assets at a wide variety of incidents, from automobile accidents, train crashes, boat accidents, airplane crashes, elevator emergencies, scaffold and other high-angle incidents, confined space rescues, trench cave-ins, building collapses, tornadoes, hurricanes, earthquakes, water rescues, hazardous materials incidents . . ."

BULLOCK

14'-0"

Skokie Blvd
LIN-MAR
4172
MACK
LIN-MAR
60TON ROTATOR

LIN-MAR
LIN-MAR
877 965 4200

Chicago f refighters look in a warehouse for hidden fire.

Chicago Fire Department Squad Co. 1 at the scene of a car that crashed into a gas meter.

Chicago Fire Chief (Ret.) Mike Burns at the scene where a van exploded with one person inside on Dec. 31, 2008.

Chicago firefighters decontaminate hazmat technicians after leaving the hot zone on Jan. 5, 2008.

Glenview, IL, Fire Department responds to a rollover crash to extricate multiple people on Dec. 18, 2011. Everyone survived.

Chicago firefighters stabilize a vehicle that went into a building and trapped the driver.

Chicago Fire Department hazmat team and Squad 2 neutralize acid poured on a vehicle on July 28, 2008.

A hazmat scene where someone poured acid on a car in 2008.

Chicago Fire Department Squad 5 and Truck 18 members assist a tree trimmer who got stuck in a tree on June 2, 2019.

An acid Leak at a Chicago chemical plant on Jan. 1, 2008.

Mabas Division 3 Technical Tescue Team responded to a scene where a man working in a hole was trapped in Niles, IL, on July 19, 2014. The man survived.

Chicago Fire Department Squad 5 tends to a memorial wreath, the scene of a fatal collision of a bus and fire truck that killed Firefighter Billy Grant.

Chicago Fire Department Squad 5A Snorkel extinguishes a train station fire on May 27, 2017.

Chicago Fire Department crash rig from the airport uses purple K to extinguish an electrical vault fire on May 14, 2020. Purple K is a dry chemical that's effective on pressurized gases and flammable liquids.

Deputy Fire Commissioner (Ret.) Tim Sampey on the scene on May 14, 2020

A freight train collision fire in Bartlett, IL, on Nov. 3, 2011

Fire chiefs stand by while workers separate train cars after a collision on Nov. 3, 2011.

Harvey, IL, Fire Captain **Keith Vallow** said Tim Olk's professionalism sets him apart. The first time they met, Olk asked him about the rules of the fireground, a suburban fire department south of Chicago. "Timmy has been helpful in getting photos if there's an investigation after a fire."

One of Olk's greatest assets is showing respect to families who have lost loved ones. Olk took photos at Vallow's father's funeral.

Harvey, IL, Fire Department works a fire at the abandoned Dixie Square Mall, where the 1980 film *The Blues Brothers* was shot.

Boone County, IL, firefighters cut through metal siding to remove burning corn on Oct. 11, 2017.

Boone County, IL, Fire District 2 responds to an anhydrous ammonia leak at a farm on Aug. 20, 2022.

A Boone County, IL, Fire District 2 deputy chief assesses the scene.

Gary, IL, Fire Department Assistant Chief Rob Grady stands by at an extra-alarm fire while crews set up a deluge truck with foam on Nov. 11, 2020.

R8

53'
MSA
MICHAEL

CHICAGO FIRE
CHICAG
SQUAD CO
1A

ONE WAY
GROUP INC

NO LEFT TURN
ONE WAY
1203

ONE WAY
CHICAGO FIRE DEPT
CHICAGO FIRE DEPT
LIEUTENANT

# GROUP SHOTS

The dangers faced by Chicago Fire Department Truck Company 20 following a still-and-box alarm fire where a roof ladder burns and flames shoot through a roof.

Boone County, IL, Fire District 2 Firefighter Hauser

Boone County, IL, Fire District 2 Lt. Josh Dummer and Deputy Chief Cory Dovenmuehle

Boone County, IL, Fire District 2 photographer Nikki Dovenmuehle

Boone County, IL, Fire District 2 Firefighter Kevin Schultz responds to a rollover crash.

Chicago Fire Department Squad Company 5 second shift after a southside fire

Chicago Fire Department Engine Company 116 and Squad Co. 5 after a fire

Chicago Fire Department Lt. Jay Bird. Engine 101

Chicago Fire Department 18th Battalion

Chicago Fire Department Squad 1, first shift, after a fire

Chicago Fire Department Squad Company 1, first shift, following a still-and-box alarm fire

Chicago Fire Department Battalion 6 Chief Jake Jakubec

Chicago Fire Department Engineer Teddy Johnson with his nephew Mickey Johnson on June 30, 2019

Chicago Fire Department Engine 109 Firefighter and cook Eddie Negron

Chicago Fire Department Lt. Art Aranda and Chicago Squad 1 Captain Sal Mastrodominico

Chicago Fire Department Firefighter Larry McCormack

Chicago Fire Department Engine 109 and Truck 32, first shift

Chicago Fire Department Firefighter Chris Olszewski, Firefighter Kelly Burns, and Firefighter Clay Whiteley

Chicago Fire Department Engine Company 117 at a 2-11 alarm fire on Jan. 7, 2017

Chicago Fire Department Squad 2 discovered coal dust between floors after they pulled ceilings at an extra-alarm apartment building. Everyone cleaned off with Fire Wipes on Jan. 2, 2020.

Chicago Fire Department Engine 95 Truck 26 at a vacant three-story, cutting metal off windows on Oct. 31, 2019

Squad 1 Chicago firefighters with Gary Fire Department Firefighter Tony Buckrop, Indianapolis Firefighter Mike Estridge, and Battalion 1 Chief Steve Little

Chicago Fire Department Battalion Chief Mitch Crooker and fan Talon Joynt

Chicago Fire Department Chiefs Tom Carbonneau, Mitch Crooker, John Collins, and Squad 1 Captain Sal Mastrodominico

Chicago Fire Department Battalion Chief Mitch Crooker and Chaplain Tom Mulcrone with Chief Tom Carbonneau

Chicago Fire Department Battalion 18 Chief Tim Sampey with his father, 9th Battalion Chief Robert Sampey, and Battalion 9 Chief Jim O'Donnell on June, 6, 2020.

Chicago Fire Department Deputy District Chief (Ret.) Steve Chikerotis (right) and *Chicago Fire* actor Taylor Kinney

*Chicago Fire* television actors with Chicago Fire Department Deputy District Chief (Ret.) Steve Chikerotis

*Chicago Fire* actors David Eigenberg ("Hermann") and Joe Minoso ("Cruz") with actual firefighters from Squad 5: John Glennon and Squad 2 Firefighter Bryan Velez, on set

Des Plaines, IL, Fire Department New Squad 63 dedication

Des Plaines, IL, Fire Department Lt. Klikas and Lt. Milton at the New Squad 63 dedication

Des Plaines, IL, Fire Department second Shift New Squad Dedication

Sandwich, IL, Fire Department after a funeral for Fire Chief Jason "Jake" Pruski on Aug. 3, 2019

Nipsta, IL, Fire Academy instructors from left: Tim Smeltzer, Mike Mitchell, Jim Pack, Mike Bronk, Pat Schey, and Scott Exo

Streamwood, IL, Fire Department

Chicago Fire Department Captain Meg Alheim and Nipsta, IL, Fire Academy graduates

Valparaiso, IN, Fire and Police photo shoot